Understanding the Elements of the Periodic Table™

SULFUR

Aubrey Stimola

New York

For my mother, the incredible Rosemary Stimola, who still knows when I need to be reminded about the pale green pants with nobody inside them

Published in 2008 by The Rosen Publishing Group, Inc.
29 East 21st Street, New York, NY 10010

First Edition

Library of Congress Cataloging-in-Publication Data

Stimola, Aubrey.
Sulfur/Aubrey Stimola.—1st ed.
p. cm.—(Understanding the elements of the periodic table)
Includes bibliographical references and index.
ISBN-13: 978-1-4042-1961-8
ISBN-10: 1-4042-1961-7
1. Sulfur—Popular works. 2. Periodic law—Popular works. I. Title.
QD181.S1S75 2007
546'.723—dc22

2006100309

Manufactured in China

On the cover: Sulfur's square on the periodic table of elements. Inset: The subatomic structure of a sulfur atom.

Contents

Introduction

Have you had any experience with the element sulfur? You may not be aware of it, but you most definitely have. Pure sulfur and sulfur combined with other materials can be found all around us. It is in the tires on your bicycle, some of the foods on your plate, and even on one of the moons of Jupiter. Sulfur is inside of us, too, where it plays many essential biological roles. In fact, without sulfur, most living things, including humans, would be unable to survive.

In addition to its biological significance, sulfur has many other important roles in industry, in the protection of agricultural crops, and in food preservation, just to name a few. However, despite their great value to mankind, sulfur-containing substances may also be extremely dangerous to people and to the environment. For that reason, they must be managed very carefully.

Chapter One
The Devil's Element

Sulfur, identified by the chemical symbol S, is the seventeenth most abundant element on Earth. In its most common elemental form, it is a bright-yellow, soft and crumbly substance that belongs to a category of elements known as nonmetals. As a nonmetal, it is a poor conductor of electricity and heat. Sulfur does not dissolve in water, but it does dissolve in certain organic (carbon-containing) solvents.

The name "sulfur" probably comes from the Sanskrit *sulvere*, the Latin *sulfurium*, or the Arabic *sufra*—all of which mean "yellow." While many people associate sulfur with the smell of rotten eggs, pure sulfur actually has no odor or taste.

Historically Speaking

Sulfur was known to humans long before the field of chemistry developed. In prehistoric times, a reddish-brown, sulfur-containing pigment called cinnabar was used in cave paintings and other art forms. Archaeological evidence also shows that prehistoric humans may have taken sulfur-containing tablets, probably to relieve digestive problems. As early as 2000 BCE, ancient Egyptians burned sulfur to produce fumes used to bleach fabrics. In the eighth century BCE, the Greek poet Homer referred to sulfur in his famous work, the *Odyssey*. The hero Odysseus requests

Pure sulfur is a bright-yellow, crumbly solid that has no odor. The element has many industrial, agricultural, biological, and medical uses.

sulfur so that he could burn it and "purify the house." This is one of the oldest references to the use of sulfur dioxide as a fumigant, a gas that can kill pests and vermin.

In some translations of the Bible, sulfur is referred to as brimstone, which, along with fire, destroyed the famed cities of Sodom and Gomorrah. It was this reference that may have resulted in sulfur's nickname, "the Devil's Element." During the Byzantine Empire, a fiery substance hurled at enemy ships probably contained sulfur, though the exact recipe for this "Greek fire" has since been lost. In tenth-century China, the first recipe for gunpowder called for sulfur. In fact, sulfur is still a component of gunpowder and other explosives today.

Antoine-Laurent Lavoisier *(right)* is considered one of the fathers of modern chemistry. His work helped other chemists understand the nature of elements.

It's Elemental

Despite its use throughout history, sulfur did not reveal its true chemical identity until the late 1700s. Around that time, French chemist Antoine-Laurent Lavoisier (1743–1794) placed sulfur among the elements, unique substances that could not be separated into purer substances. Lavoisier's theory became accepted scientific fact in 1809, when chemists Joseph-Louis Gay-Lussac and Louis Jacques Thénard proved Lavoisier correct. But what is an element? And what makes an element like sulfur so unique?

An element is a substance made up of only one kind of atom. Atoms are the basic building blocks of all the matter in the universe; they cannot be broken down into simpler substances. Pure sulfur is made up of only sulfur atoms. Similarly, pure carbon is made up entirely of carbon atoms. So far, scientists have discovered or created 116 different kinds of atoms, each of which represents a unique element. Scientists can tell atoms apart by observing their physical and chemical properties.

From Atoms to Molecules and Compounds

Like individual ingredients in a recipe, elements combine with other elements to create the variety of substances all around you. Specifically, atoms of

Sulfur's square on the periodic table. Similar to all other elements, pure sulfur is made up of only one type of atom.

the various elements combine to create molecules and compounds. Molecules contain at least two atoms, and compounds contain at least two different kinds of atoms. For example, two atoms of the element hydrogen (H) combine with one atom of the element oxygen (O) to create a molecule of water (H_2O). Later, we will see how atoms of sulfur combine with atoms of other elements to make some useful compounds. First, however, we must learn about the unique properties of sulfur itself.

Chapter Two
A Closer Look at Sulfur

Sulfur is a solid at room temperature. This is due to the unique arrangement of sulfur atoms. Rather than existing all alone or in pairs like many other types of nonmetal atoms, sulfur atoms naturally arrange themselves in crooked rings of eight sulfur atoms each. These rings, or S8 molecules, are often referred to as "crowns" for reasons that are obvious from diagrams that illustrate the structure (see page 10). Multiple S8 crowns will naturally arrange themselves in organized crystal patterns. These distinct arrangements of S8 molecules result in the allotropes, or different forms, of sulfur.

Allotropes of Sulfur

The most common naturally occurring allotrope of elemental sulfur is orthorhombic sulfur, which is a brittle yellow solid. At temperatures above 205 degrees Fahrenheit (96 degrees Celsius), however, another crystal arrangement forms. This one, known as monoclinic sulfur, is a paler yellow. A third allotrope of sulfur named amorphous, or plastic sulfur, occurs when molten (melted) sulfur is cooled so quickly that its atoms do not have time to arrange into S_8 crowns. Unless the atoms are in S_8 crowns, orthorhombic or monoclinic crystal arrangements cannot form. Instead,

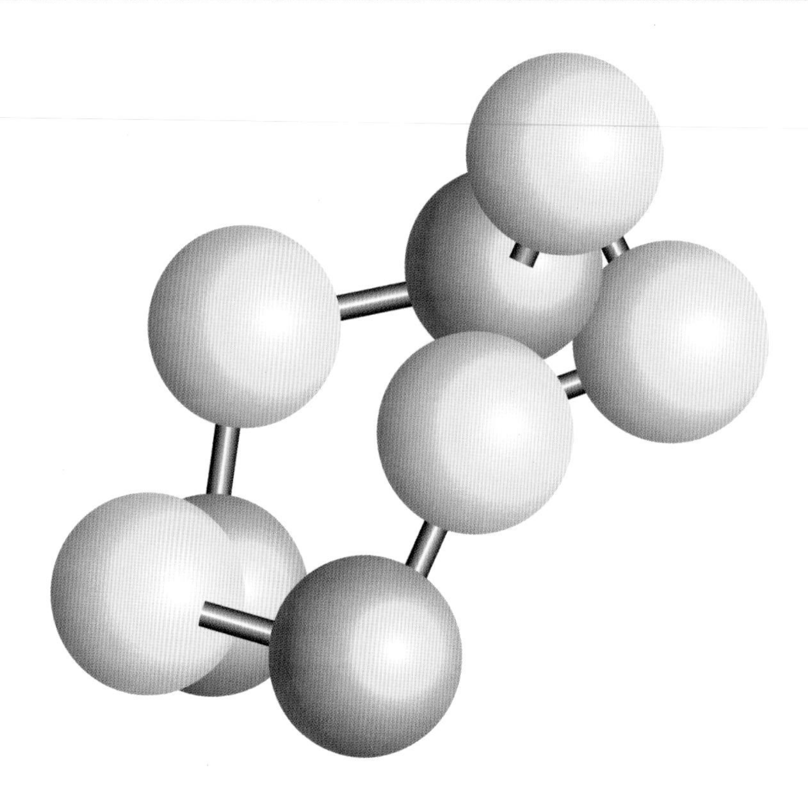

The S_8 molecule, also called a sulfur crown, is made up of eight individual sulfur atoms. Multiple S_8 crowns combine to create different forms of pure sulfur.

the quickly cooled individual sulfur atoms form long chains that give plastic sulfur its characteristic softness and elasticity, or stretchiness.

Breaking the Rules: An Unusual Property of Sulfur

The term "viscosity" refers to a substance's resistance to flow. Honey, for example, has a higher viscosity than water, as anyone who has tried to pour it out of a jar can tell you. Typically, when substances are melted, they become less viscous. In other words, as temperature rises, viscosity tends to decrease. If you were to heat up honey, you would find that it is

Sulfur Snapshot

16 S 32

Chemical Symbol:	S
Classification:	Nonmetal
Properties:	Pale yellow, odorless, brittle solid, insoluble in water
Group Number:	16 (also known as the chalcogens)
Discovered by:	Identified as an element by Antoine-Laurent Lavoisier, c. 1770s
Atomic Number:	16
Atomic Weight:	32.06 atomic mass units (amu)
Electrons:	16
Density at 68°F (20°C):	(Orthorhombic Allotrope) 2.07 g/cm^3 (Monoclinic Allotrope) 1.96 g/cm^3
Melting Point of Orthorhombic Allotrope:	235°F (112.8°C)
Boiling Point of Orthorhombic Allotrope:	832.3°F (444.6°C)
Commonly Found:	Earth's crust, soil, seawater, the atmosphere, meteorites, volcanoes, hot springs, swamps, fossil fuels, other planets, and bound to other metals

These spaghetti-like strands of plastic sulfur resulted from pouring molten sulfur into cold water. The rapid cooling prevents the reformation of S_8 crowns.

much easier to pour out of its container. Sulfur, however, does not play by these rules. Just above its melting point of 235°F (112.8°C), orthorhombic sulfur is a flowing, orange-brown liquid. As the temperature rises, though, the molten sulfur actually becomes more viscous. The reason for this is that heating the sulfur above the melting point causes the S_8 crowns to break. The individual sulfur atoms then rearrange themselves into long chains called polymers. These long chains become tangled and make molten sulfur very resistant to flow. But if the temperature increases beyond a certain point, these sturdy polymer chains, too, will break apart, resulting in black, runny, liquid sulfur.

Chapter Three
Subatomic Particles

All atoms, including sulfur atoms, are made up of smaller, subatomic particles called protons, neutrons, and electrons. Knowing the number of subatomic particles that an atom has allows us to determine what kind of atom it is. For example, sulfur—and only sulfur—has sixteen protons. The number of protons in an atom is also the atom's atomic number. Therefore, the atomic number of sulfur is 16. If an atom has an atomic number other than 16, it is not a sulfur atom.

Protons, Neutrons, and Electrons

Protons, which have a positive electrical charge, are found in an atom's center, or nucleus. Also contained within the nucleus are an atom's neutrons. Neutrons are neutral, meaning they have no electrical charge, so the nucleus of an atom is always positively charged. Typically, sulfur has sixteen neutrons within its nucleus. However, unlike the number of protons, which always remains the same, the number of neutrons can vary. Different types of sulfur atoms are known as isotopes of sulfur. Sulfur has dozens of isotopes, but only those with sixteen, seventeen, eighteen, and twenty neutrons are stable. Stable isotopes do not break down. Sulfur's unstable isotopes, on the other hand, will break down into other sulfur isotopes in order to regain stability—a property known as radioactivity.

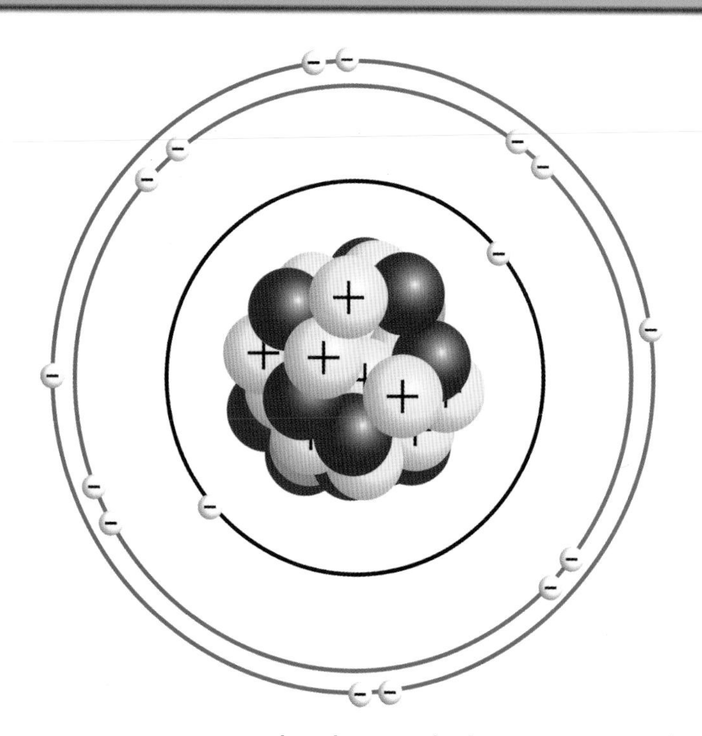

Diagram of a typical sulfur atom. The nucleus contains sixteen positively charged protons *(yellow)* and sixteen neutrons *(black)*, which have no charge. Sixteen negatively charged electrons *(light blue)* travel around the nucleus.

Negatively charged electrons circulate around the positively charged nucleus in spaces called electron shells. Each electron shell can hold only a certain number of electrons, with additional electrons spilling over to the next available electron shell. The number of electrons in a neutral atom always matches the number of protons, or atomic number. Can you determine, then, how many electrons a sulfur atom has? If you said sixteen, you're right. Sulfur's atomic number (16) tells you that a sulfur atom has sixteen electrons. The protons and neutrons in an atomic nucleus have about the same mass; electrons are much lighter.

Electrons play two very important roles. First, the negatively charged electrons are attracted to the positively charged protons in the nucleus. This attraction, electromagnetic force, is what holds the atom together. Because the number of electrons and protons is equal, the negative and positive charges cancel each other out. This leaves the atom with no overall charge. Second, an atom's electrons are central to its ability to bind to other atoms and to form compounds. Particularly important are the electrons in the outermost electron shell, known as valence electrons.

Electrons and Reactivity

An atom tends to lose, gain, or even share electrons in order to fill its outermost electron shell. When an atom's outermost electron shell is full, the atom is more stable and less reactive.

A typical sulfur atom has three electron shells; the first contains two electrons, the second contains eight electrons, and the third contains six electrons. This third electron shell is stable when it contains eight electrons. So, to attain stability, the sulfur atom will acquire two more electrons. It can do so in two ways: either it will take two electrons from an atom that gives them up, or it will share two electrons with another atom. Taking electrons results in the formation of ions, producing an ionic bond. Sharing electrons with another atom does not produce ions and instead results in a covalent bond.

Ionic Bonding

When an atom gains electrons, which carry a negative charge, the atom's electrons and protons no longer cancel each other out. Instead, for every electron an atom gains, its negative charge increases. Similarly, for every electron an atom loses, its positive charge increases. Charged atoms are called ions. Two sodium (Na) atoms will each give up one electron to a single sulfur atom to form a compound made of two stable sodium ions with a positive one charge ($2Na^{1+}$) and one stable sulfur ion with a negative two charge (S^{2-}). These ions are stable because both the Na atoms and the S atom now have filled their outermost electron shells. Because the S^{2-} ion has a charge that is equal and opposite to the combined charge of the two Na^{1+} ions ($1^+ + 1^+ = 2^+$), all three ions "stick" together in a stable compound called a sulfide.

$$2Na + \mathbf{S} \longrightarrow 2Na^{1+} + \mathbf{S^{2-}} \longrightarrow Na_2\mathbf{S}$$

Sulfides, Sulfates, and Sulfites: What Are the Differences?

The sulfide ion, S^{2-}, is a sulfur atom that has gained two electrons. In addition to the sulfide ion, there are other common sulfur-containing ions. Some of them involve a sulfur atom, a particular number of oxygen atoms, and a negative two charge. There is the sulfate ion, SO_4^{2-}, which combines with other elements to form compounds collectively known as sulfates. A common sulfate is barium sulfate, $BaSO_4$.

Another common sulfur ion is the sulfite ion, SO_3^{2-}, which has one less oxygen atom than the sulfate ion. The sulfite ion combines with other elements to create compounds known as sulfites, such as sodium sulfite, Na_2SO_3. Sulfides, sulfates, and sulfites each have different chemical properties.

Covalent Bonding

In certain situations atoms share electrons, rather than accepting or giving them up entirely. The sharing of electrons between atoms is known as covalent bonding. Covalent bonding does not involve ions.

Sulfur, in addition to forming ionic bonds, is capable of forming covalent bonds. For example, two hydrogen atoms each share an electron with one sulfur atom to create hydrogen sulfide.

Just as in ionic bonding, all of the atoms involved in covalent bonding become more stable. Sulfur has the ability to bond, ionically or covalently, with several other elements to create very important compounds. Sulfur forms ionic bonds when it combines with metal atoms and covalent bonds when it combines with nonmetal atoms. In chapter 5, we will take a closer look at some of the more important of these compounds. First, however, we'll learn more about sulfur's relationship to the other 115 elements.

Chapter Four
Sulfur and the Periodic Table of Elements

Sulfur is one of 116 elements, ninety of which occur naturally. The others have been created in laboratories under conditions that don't normally occur. It is very possible that there are elements yet to be discovered.

Taken together, the elements are very complex. So, it is helpful to organize them in a convenient and meaningful way. In 1869, Russian chemist Dmitry Mendeleyev attempted to do just that. First, Mendeleyev arranged the elements in horizontal rows in order of increasing atomic weight. (Atomic weight is a very specific number that tells you the average mass of an atom of a particular element. It is approximately the sum of the number of protons and neutrons in the nucleus.) After arranging the known elements by atomic weight, he noted regular, repeating patterns of such properties as density, reactivity, and boiling point. These repeating patterns—or periods—led Mendeleyev to call his construction the periodic table of elements.

Mendeleyev's arrangement of the elements was not perfect, however, as there were gaps between elements. Rather than start from scratch, the ambitious chemist asserted that the gaps in his table represented elements that had yet to be discovered. Based on the location of these gaps, Mendeleyev even went so far as to predict the properties of the missing elements. Within twenty years of the first publication of Mendeleyev's

Russian scientist Dmitry Mendeleyev *(left)* created the first popular periodic table of elements.

periodic table, three of these gaps were filled in with newly discovered elements. Incredibly, the characteristics closely matched Mendeleyev's predictions!

More than 100 years have passed since the publication of the first periodic table of elements. In that time, the table has undergone several changes, including the addition of more than fifty elements. But Mendeleyev's overall scheme still serves as its basis. Today, the elements are arranged in order of increasing atomic number, or number of protons, rather than atomic weight. They are arranged in horizontal rows, or periods, labeled 1 through 7. Moving from left to right across a period, each successive element has one more electron in its outer electron shell than the one before it. This leads to a predictable pattern in the physical and chemical behaviors of the elements.

The eighteen vertical columns of the periodic table are called groups. In one system, groups are labeled IA through VIIA, IB through VIIIB, and 0. In a second system, groups are labeled 1 through 18. (See periodic table on pages 40–41.) Elements in a group have the same number of electrons in their outer shells, resulting in some similar chemical behaviors. For this reason, the elements of a given group are considered members of the same family.

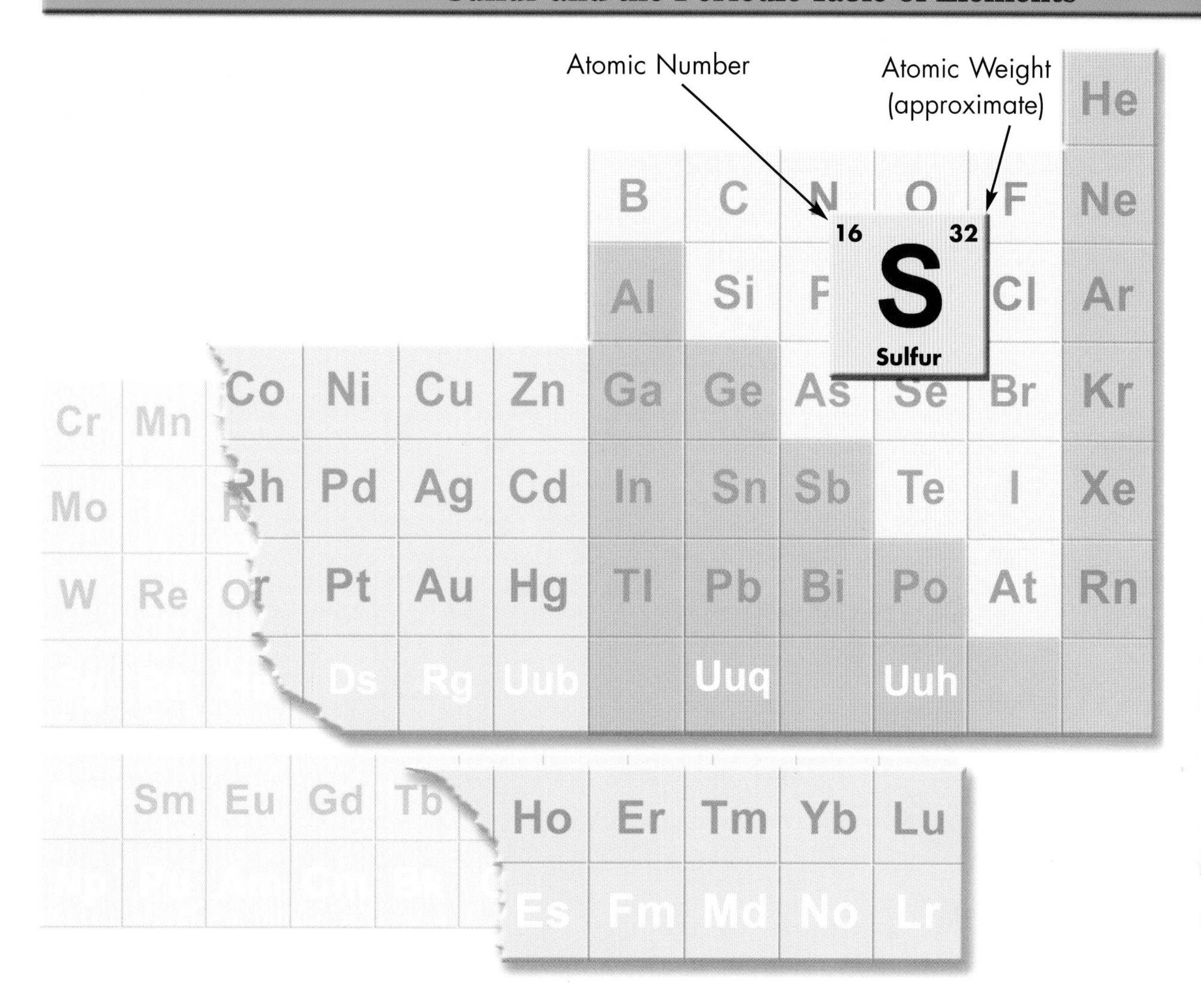

Sulfur's group on the periodic table is sometimes called the chalcogen group. Other elements in sulfur's group include oxygen (O), selenium (Se), tellurium (Te), polonium, (Po), and ununhexium (Uuh).

Sulfur and the Periodic Table

Sulfur is located in Period 3 and Group VIA (16) of the periodic table. All elements in group VIA have six electrons in their outer shells. This means that a sulfur atom gains two electrons from other atoms to become stable.

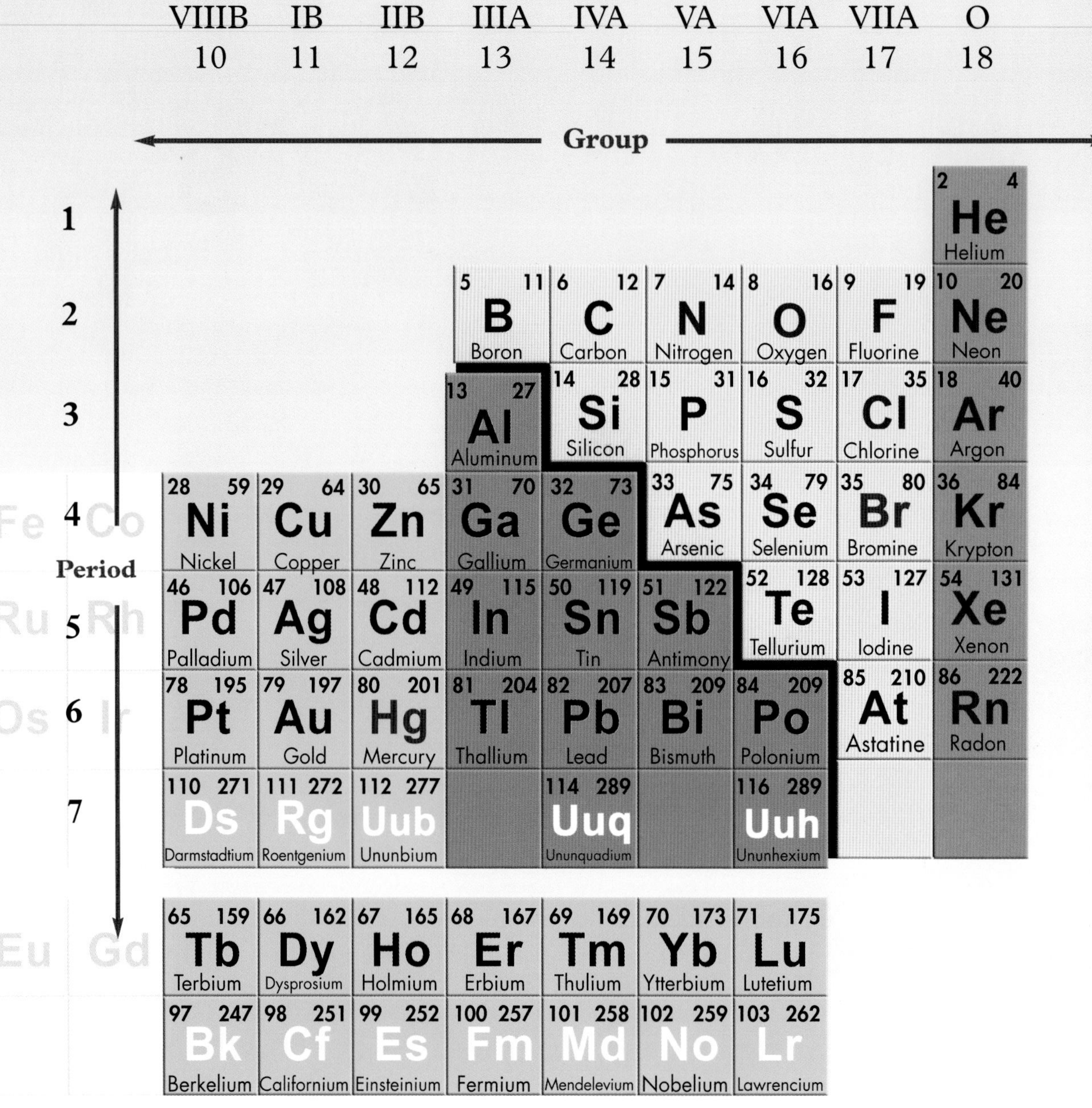

The periodic table is divided into columns (groups or families) and rows (periods). This portion of the table shows only groups 10 through 18. The elements in yellow squares to the right of the bold black staircase line—including sulfur—are known as nonmetals.

This strong tendency to gain electrons makes sulfur and the other Group VIA elements very reactive.

The number in the upper left-hand corner of sulfur's square represents the element's atomic number. Because all sulfur atoms have sixteen protons, sulfur's atomic number is 16. The number in the upper right-hand corner represents the element's approximate atomic weight.

Armed with some background information on sulfur, how it reacts with other atoms, and where it fits in relation to other elements, we can take a look at some of the important uses of sulfur and its compounds. They are quite varied.

Chapter Five
Sulfur Compounds and Their Uses

Sulfur and sulfur compounds are used in the production of some medications, wine, gunpowder, matches, and artificial sweeteners. They are used in pesticides and products that kill fungi, weeds, and rodents. Sulfur is also present in grease- and oil-removing detergents, in some types of batteries, and in fertilizers. The element is necessary for the vulcanization (hardening) of rubber and in some food preservation techniques.

One of sulfur's most important uses is in the manufacturing of sulfuric acid, a chemical with countless industrial roles. Despite their usefulness, however, sulfur compounds can be extremely dangerous, toxic, and harmful to the environment. For this reason, they must be handled with great care.

Sulfuric Acid—H_2SO_4

The primary use of sulfur is to manufacture sulfuric acid. Discovered by an eighth-century alchemist, this clear, odorless, and extremely corrosive oily acid was originally called oil of vitriol.

Mixing specific amounts of sulfuric acid with water produces different concentrations of the acid and, therefore, different strengths. This process is called dilution. Different strengths of sulfuric acid have different uses. For example, 10 percent sulfuric acid (90 percent water) is commonly

A Stinky Substance?

Have you ever heard that sulfur smells like rotten eggs? Actually, pure sulfur has no odor at all. Many sulfur compounds, however, do have foul odors. These compounds include chemicals known as thiols, mercaptans, and disulfides. Sulfur compounds are responsible for the pungent odors associated with skunks, matches, onions, garlic—and rotten eggs. They are also responsible for the stench of pollutants from some power plants and factories. In fact, smelly mercaptans are purposefully added to natural gas to make leaks more easily detectable.

When threatened, skunks will spray a potent sulfur-containing liquid called N-bulymercaptan. They can hit a target up to fifteen feet away, so stand clear!

used in laboratories, while car batteries typically use 35 percent sulfuric acid (65 percent water).

One must be extremely careful when diluting sulfuric acid. Mixing water and sulfuric acid releases enough heat to make the container hot to the touch. In fact, when water is poured into concentrated sulfuric acid, the water immediately begins to boil and will splash out of the container. For this reason, water is never added to sulfuric acid. Instead, the acid is added in small amounts to the water.

Table sugar *(1)* contains carbon and water. After sulfuric acid is introduced *(2)*, a reaction takes place *(3 and 4)*. During the reaction, the sulfuric acid pulls water molecules out of the sugar, leaving behind a black pillar of pure carbon *(5)*.

Given its countless industrial uses, more sulfuric acid is produced annually than any other chemical. Sulfuric acid is added to phosphate-containing rocks to produce phosphate fertilizers for crops and other plants. It is also used to remove impurities from oil and mineral ores, in the creation of other important chemicals, and in wastewater processing. Sulfuric acid is involved in the production of certain paints and dyes, car batteries, and synthetic fibers like rayon. It is also crucial in the "pickling" of iron and steel, a surface-cleaning process that prepares these metals for further refining and coating, or plating, with other metals. The iron and steel are bathed in sulfuric acid, which dissolves rust and other impurities that build up on their surfaces.

Because it is highly reactive, sulfuric acid can be an extremely dangerous chemical. Even dilute (weak) sulfuric acid is capable of dissolving many metals. For example, a strip of zinc placed in a beaker of sulfuric acid will immediately begin bubbling and fizzing. In a short amount of time, no trace of the metal strip will remain. Sulfuric acid also reacts with human skin, pulling all water out of the skin cells and leaving behind an extremely painful acid burn. To protect themselves from accidental injury, scientists who work with sulfuric acid always wear protective goggles and clothing.

Sulfur Dioxide—SO_2

Sulfur dioxide is a colorless, poisonous gas used in the production of sulfuric acid and as a food preservative. As a preservative, sulfur dioxide is used in small quantities to deter the growth of mold and bacteria in various foods and beverages, including raisins, dried apricots, and wine.

Sulfur dioxide is also used to give electrons to compounds in certain dyes and pigments, resulting in decolorization, or bleaching. This property of sulfur dioxide makes the compound useful in the production of some fabrics and papers. Over time, however, oxygen in the air takes the donated electrons back from the bleached pigments in a process called oxidation.

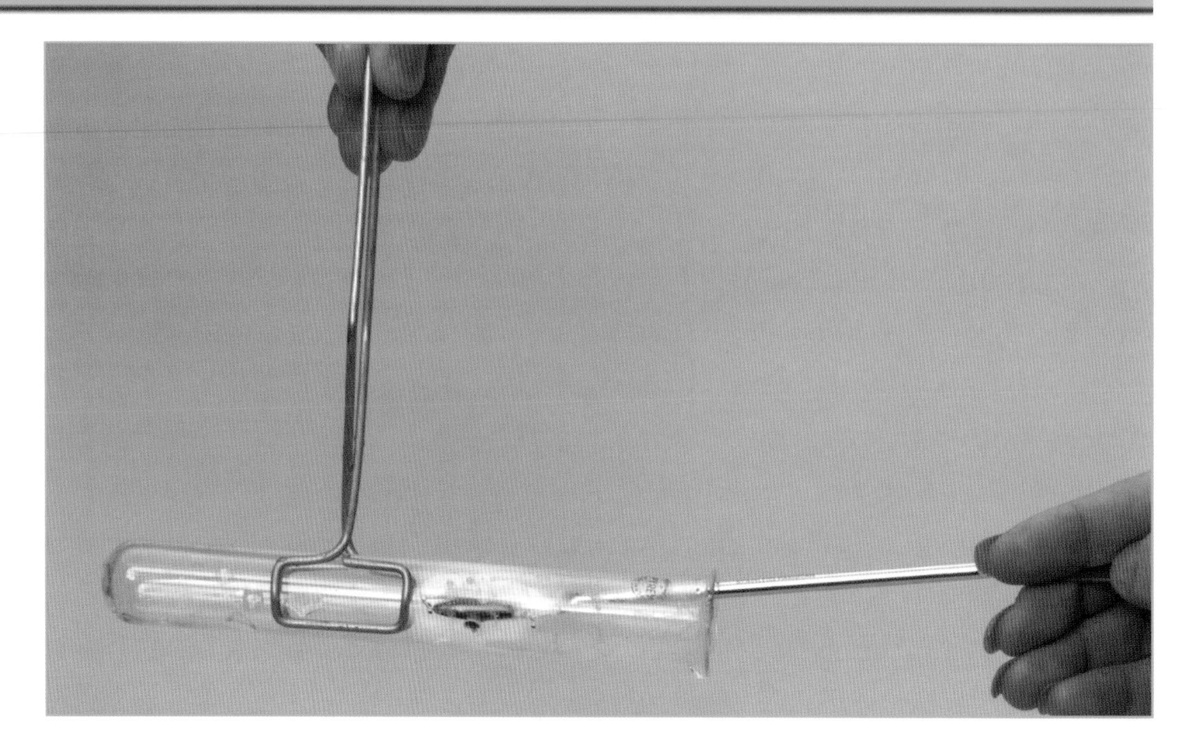

Sulfur burns in a tube of pure oxygen, producing an electric-blue flame. The reaction also produces sulfur dioxide (SO_2), the gas that accounts for the sharp odor you smell when a match is lit.

As a result of oxidation, bleached items regain some color. For example, bleached newspaper will eventually return to its original yellow color.

Although useful, sulfur dioxide can cause problems. It is a harmful by-product of many industrial and power-generating processes, particularly the combustion, or burning, of coal and petroleum. Sulfur dioxide is also emitted naturally by volcanoes and certain microorganisms, though in less dangerous amounts.

Acid Rain

Each year, millions of tons of sulfur dioxide (SO_2) are released from the smokestacks of coal-burning power plants worldwide. When sulfur dioxide is released into the air, it reacts with oxygen to create sulfur trioxide, or SO_3.

When SO_3 mixes with water droplets in the air, sulfuric acid—the very strong, corrosive acid we discussed earlier—is produced. Sulfur dioxide can also mix directly with water droplets in the air to produce a weaker acid called sulfurous acid. Both of these acids then fall to the earth as acid rain.

Acid rain is extremely costly to the environment. It damages trees at high altitudes and raises the acidity of lakes, streams, and soils, thereby disrupting the ecological balance needed to sustain many plants and animals. Acid rain is costly to humans as well. It destroys the facades of historical buildings and statues by causing them to break down faster. Many common building materials are eroded by acid rain, including limestone, sandstone, and some metals and paints.

Acid rain caused the destruction of this mountain forest, which was once home to healthy spruce and balsam trees.

Hydrogen Sulfide—H_2S

Hydrogen sulfide is a foul-smelling, flammable, poisonous gas that results from the decay, or breakdown, of organic matter. (The odor of rotten eggs is actually the odor of hydrogen sulfide.) It is also emitted from volcanoes and hot springs, and it is a component of natural gas. The powerful odor sometimes present near swamps or sewers comes from hydrogen sulfide gas. Many bacteria in these low-oxygen environments consume sulfur-containing substances in order to produce energy, and hydrogen sulfide gas is released as a by-product of this process. The hydrogen sulfide released by bacteria in the human digestive tract and mouth is, in part, responsible for the odors associated with flatulence and bad breath!

Inhaling hydrogen sulfide gas is potentially deadly. The poisonous gas affects many systems in the body, most notably the nervous system. Despite its toxicity, however, hydrogen sulfide gas has found some use in industry, particularly in the formation of other important sulfur compounds, the metal sulfides.

Metal Sulfides

Metal sulfides are compounds that contain sulfur and one or more metals. For example, the combination of lead (Pb) and sulfur results in lead sulfide, or

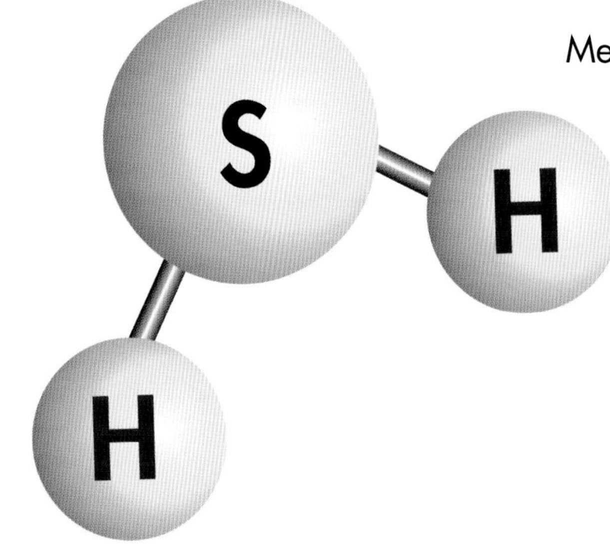

A molecule of hydrogen sulfide contains two hydrogen ions (H) bonded to one sulfur ion (S).

Coal Combustion in Electricity-Generating Plants

Many power plants in the United States and other countries burn coal to generate electricity. Because coal contains sulfur, these power plants are an enormous source of poisonous, acid rain–producing sulfur dioxide. To help manage this problem, researchers are seeking ways to reduce sulfur dioxide gas emissions from power plants. One approach is to burn only low-sulfur coals. In addition, the Environmental Protection Agency requires many coal-fired plants to use special devices called scrubbers. These are built into smokestacks to remove much of the sulfur dioxide from the smoke before it is released to the air. Other fuels, including some types of coal, are desulfurized using chemicals to bind and remove the sulfur they contain.

Electrical power plants that burn coal are a major source of sulfur dioxide around the world. Sulfur dioxide in the atmosphere produces damaging acid rain.

Fool's gold is the mineral pyrite, which is actually iron sulfide. It looks like real gold, but unlike the precious metal, this combination of sulfur and iron is nearly worthless.

PbS. One of the better-known metal sulfides is pyrite, an iron sulfide (FeS_2) nicknamed "fool's gold" due its striking resemblance to real gold ore. Many of the metal sulfides occur naturally in minerals; others are made in laboratories using the sulfur from hydrogen sulfide gas.

Sulfates

Sulfates are compounds that contain sulfur and oxygen in the form of the sulfate ion, SO_4^{2-}. The sulfate ion can ionically bond to many metal ions. Zinc sulfate ($ZnSO_4$) is a common metal sulfate. Sodium sulfate (Na_2SO_4) is a common filler in laundry detergents. Such fillers do not actually act as the cleaning agent, but they enhance the detergent's cleaning action. Sodium sulfate is also used in the textile, paper processing, and glass industries. Copper sulfate ($CuSO_4$) is a bright-blue salt that can be used as a fungicide, in hair dyes, and in leather processing. Other metal sulfates are used as catalysts to speed up chemical reactions.

Chapter Six
Other Important Uses of Sulfur

Sulfur has many more uses than we can explore within the scope of this book. Its unique atomic structure gives the element an astonishingly wide array of applications in fields from warfare to medicine.

Making Rubber Stronger

Rubber is used to make a variety of items we encounter daily, from waterproof raincoats and tire treads to pencil erasers and elastic bands. Did you know that rubber grows on trees? Well, not exactly. Natural rubber is made from latex, a substance in the sap of rubber trees. Latex is a polymer called polyisoprene, which is made up of smaller molecules joined together in a long chain. The random polymer structure of latex gives rubber some remarkable qualities. For example, it is elastic (it can be stretched repeatedly beyond its original length) and it is largely resistant to water, temperature changes, and many chemicals.

So, what does rubber have to do with sulfur? In the 1830s, Charles Goodyear discovered that heating a mixture of natural latex and sulfur turned soft, pliable rubber into a much harder substance with broader applications. The process, named vulcanization—after Vulcan, the Roman god of fire—is still used today.

How does vulcanization work? When sulfur and latex are mixed and heated, the S_8 crowns bond to the latex polymers, creating cross-links between them. The sulfur cross-links anchor the polymers in place. This makes vulcanized rubber much more heat-resistant and less likely to wear down. Vulcanized rubber is crucial to the production of car and truck tires, boots, hoses, brake and engine parts, and even hockey pucks.

An Element of War

Sulfur has long found use in war. Warriors of the Byzantine Empire hurled "Greek fire" at their enemies. This fiery liquid, which contained elemental

This page from an ancient book shows a Byzantine naval warship attacking an enemy ship with Greek fire.

sulfur, was especially useful when fighting naval battles, as it would stick to anything it touched and continued to burn when wet.

Using elemental sulfur as a principal ingredient, the Chinese invented gunpowder, the first explosive, around the tenth century. The recipe spread quickly to Europe and beyond. Today, sulfur-containing gunpowder, a low-level explosive, is still used in fireworks and signal flares. Some modern high explosives require sulfuric acid in their production.

A sulfur compound was also a component of mustard gas, a controversial and highly toxic weapon used during World War I to inflict painful skin and lung blistering, blindness, and frequently death.

Nature, Biology, and Medicine

Would you ever guess that the same element that hardens rubber, contributes to acid rain, and is part of very toxic gases is also required for life? It's true. Without sulfur, all microorganisms, plants, and animals would die.

The Sulfur Cycle

Sulfur is constantly passed back and forth through soil, the atmosphere, and the oceans in a complex process known as the sulfur cycle. Plants, animals, and bacteria each play an important part in this cycle by absorbing sulfur in different forms and releasing it for the next phase of the cycle. Very simply put, animals eat plants that contain sulfur compounds. When the animals die, bacteria consume their flesh, turning the bodies' organic sulfur into inorganic sulfides, sulfites, and eventually sulfates. Plants then absorb these inorganic sulfur compounds to produce organic sulfur compounds that animals may once again eat.

A Dietary Element

Sulfur is found in every cell in our bodies. Without it, the process by which our cells create energy would shut down. Also, sulfur is part of some essential amino acids, the building blocks of proteins that are crucial to the cellular structures and functions that sustain life. One of these essential amino acids is methionine. In our bodies, methionine is turned into another sulfur-containing amino acid called cysteine. The sulfur in cysteine molecules plays a crucial role in folding proteins into their correct shapes. Examples of these sulfur-dependent proteins are keratin (part of hair, skin, and nails) and collagen (part of connective tissue like cartilage).

Where do we get the elemental sulfur we need to make proteins? Plants have the ability to take sulfur in ionic form (inorganic) and break it down to sulfur in molecular (organic) form. However, unlike plants, animals—including humans—cannot do this. Neither can we produce our own methionine. So, we must rely on the foods we eat to provide us with essential amino acids. Foods rich in methionine include fruits, meats, fish, nuts, soy products, mushrooms, potatoes, some beans, and such vegetables as spinach, peas, corn, cauliflower, sprouts, and broccoli. Other forms of dietary sulfur are present in milk, onions, garlic, cabbage, turnips, and brussels sprouts. As you might guess, by eating a normal, healthy diet, humans get plenty of dietary sulfur without even thinking about it.

Thiamine (vitamin B_1) is a sulfur-containing vitamin that plays an important role in helping the body convert carbohydrates and fat into energy. It is present in high concentrations in foods such as tuna fish, sunflower seeds, many cereals, and beans. Thiamine is necessary for normal growth and development, and it helps to maintain proper functioning of the heart and the nervous and digestive systems. Biotin (vitamin B_7), another vitamin crucial to energy production, also contains sulfur.

The Medical Element

Sulfur was used in prehistoric times to ease stomach upset by stimulating the digestive tract. Sometimes the element was mixed with wool fat (lanolin) to treat skin infections. While these types of medications are no longer used in their original formulations, sulfur still plays a role in

Doctors use a sulfur compound, barium sulfate, to view parts of the digestive system. The barium coats the digestive tract and shows up clearly on an X-ray.

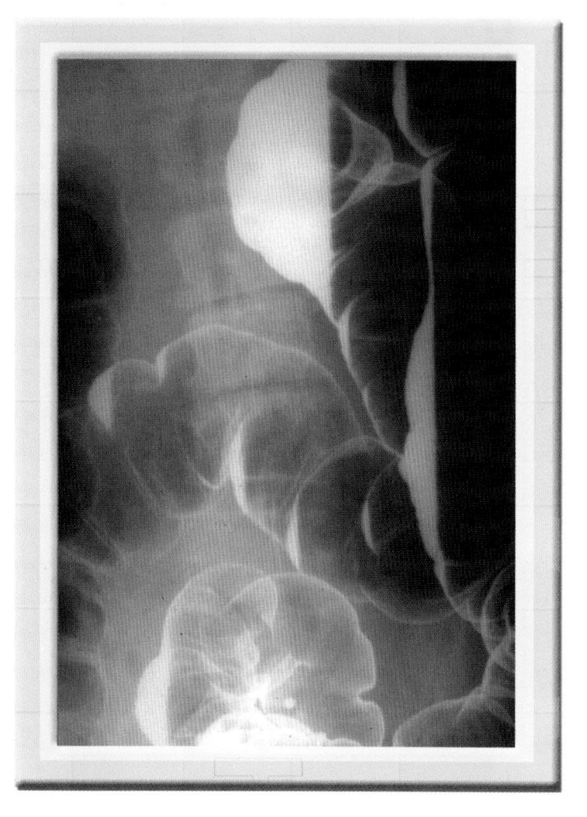

modern medicine. In the early twentieth century, substances used in some dyes were found to fight bacterial infections. Some of these substances, called sulfa drugs, are still used today. Others have been replaced with members of another sulfur-containing class of antibiotics known as the penicillins.

Dermatologists prescribe sulfur-based creams and ointments to treat acne, skin inflammation, and scabies—a parasitic infection. Some sulfates, too, are used for medicinal purposes. Magnesium sulfate ($MgSO_4$), commonly called Epsom salts, is sometimes used to slow contractions in pregnant women who go into early labor. Barium sulfate ($BaSO_4$) allows patients to safely swallow barium so that X-ray images can be taken of the digestive tract to look for blockages or other abnormalities. Barium is normally toxic to humans, but the body cannot absorb the sulfate portions of the compound, so it is safe to use. The large nuclei of the barium atoms easily absorb X-rays to create clear images on X-ray film.

Chapter Seven
Where Is Sulfur Naturally Found?

How and where do we obtain sulfur? Sulfur is a major component of soil, and small amounts are in seawater and in Earth's atmosphere. Though the amount cannot be measured directly, scientists believe that about 5 percent of Earth's core is made up of sulfur.

Volcanic Sulfur Deposits

Sulfur occurs naturally near volcanoes. Hydrogen sulfide gas bubbling up from underground is transformed into elemental sulfur by oxygen in the air. Through the 1800s, the biggest supply of sulfur in the world came from sulfur mines on the island of Sicily, home to the volcano Mount Etna. The sulfur-containing rocks collected from the mines were spread on top of slabs and set on fire. The melted liquid sulfur ran out the sides, where it was collected. This was a rather inefficient process, as more than half the sulfur was wasted. Today, similar volcanic sulfur deposits in Indonesia, Chile, and Japan serve as sources of sulfur.

Underground Sulfur Deposits

In the 1860s, speculators searching for oil stumbled across large underground deposits of relatively pure sulfur in Texas and Louisiana.

This Indonesian miner has the extremely dangerous job of collecting sulfur near an active volcano. In addition to the searing heat, the volcano's hydrogen sulfide fumes are poisonous.

Unfortunately, it was impossible with the technology of the time to affordably retrieve this high-quality sulfur. Then, in the 1890s, a German-born American chemist named Herman Frasch realized that the sulfur could be melted and pumped to the surface as a liquid. To do this, Frasch developed a process that pumped superheated steam and compressed air into the sulfur-filled underground caverns. The steam melted the sulfur, and the liquid sulfur was then forced back to the surface by the compressed air. This technique, aptly named the Frasch process, is still used. In fact, the majority of sulfur produced today is obtained in this manner.

Metal-Bound and Mineral-Bound Sulfur

While sulfur does occur in its pure elemental form, it is often bound to metals like iron. In some places, iron sulfides are still the primary source of sulfur. Other metal sulfides from which pure sulfur can be extracted are mercury sulfide (HgS, or cinnabar), lead sulfide (PbS, or galena), and zinc sulfide (ZnS, or sphalerite).

Sulfur is found in oxygen-containing sulfate minerals, including the calcium sulfates called anhydrite and gypsum. However, the process by which sulfur is extracted from sulfate minerals requires a lot of energy and wastes much of the sulfur.

Sulfur from Gases

Fuel standards designed to protect the environment from harmful sulfur gases frequently require that sulfur be removed from coal and natural gas. The reclaimed sulfur can then be put to use in any one of the many sulfur-dependent industrial processes. The hydrogen sulfide gas removed from natural gases is a tremendous source of sulfur. The natural gas is sprayed with a special chemical that causes the hydrogen sulfide to dissolve out. The isolated hydrogen sulfide is then reacted with oxygen to produce water and pure sulfur.

Using scrubbers, as much as 95 percent of the sulfur can be reclaimed from the sulfur dioxide gas emitted by coal-fired power plants and industrial plants. Scrubbers are required for newly built power plants, but U.S. environmental laws allow many older power facilities to operate without them.

Conclusion: Summing Up Sulfur

Are you surprised to find that you have had so much experience with sulfur in your daily life? Perhaps what's most amazing about sulfur is its ability

Non-Earthly Sulfur

Sulfur has been found in meteorites, on Mars, and on our moon. One of the more notable sulfur-rich places in our solar system is Io, one of Jupiter's moons. Io is covered with molten sulfur, which also erupts from the moon's many volcanoes. Bacterial life has been found in extremely sulfur-rich environments on Earth, such as deep-sea hydrothermal vents. For this reason, some scientists theorize that Io, too, may be home to sulfur-using bacteria!

NASA's Voyager spacecraft allowed experts to make exciting discoveries about Io, one of Jupiter's moons *(right)*. Photos taken on the Voyager mission showed that Io's surface has active sulfurous volcanoes, lakes of molten sulfur, and frozen sulfur dioxide.

to do both harm and good, depending on how it is used and in which compounds it is found. It is part of what makes tires durable and skunks smelly. It is vital for life, but it also can be dangerous. As you continue throughout your day, use your knowledge of sulfur to gain a new appreciation of this versatile element.

The Periodic Table of Elements

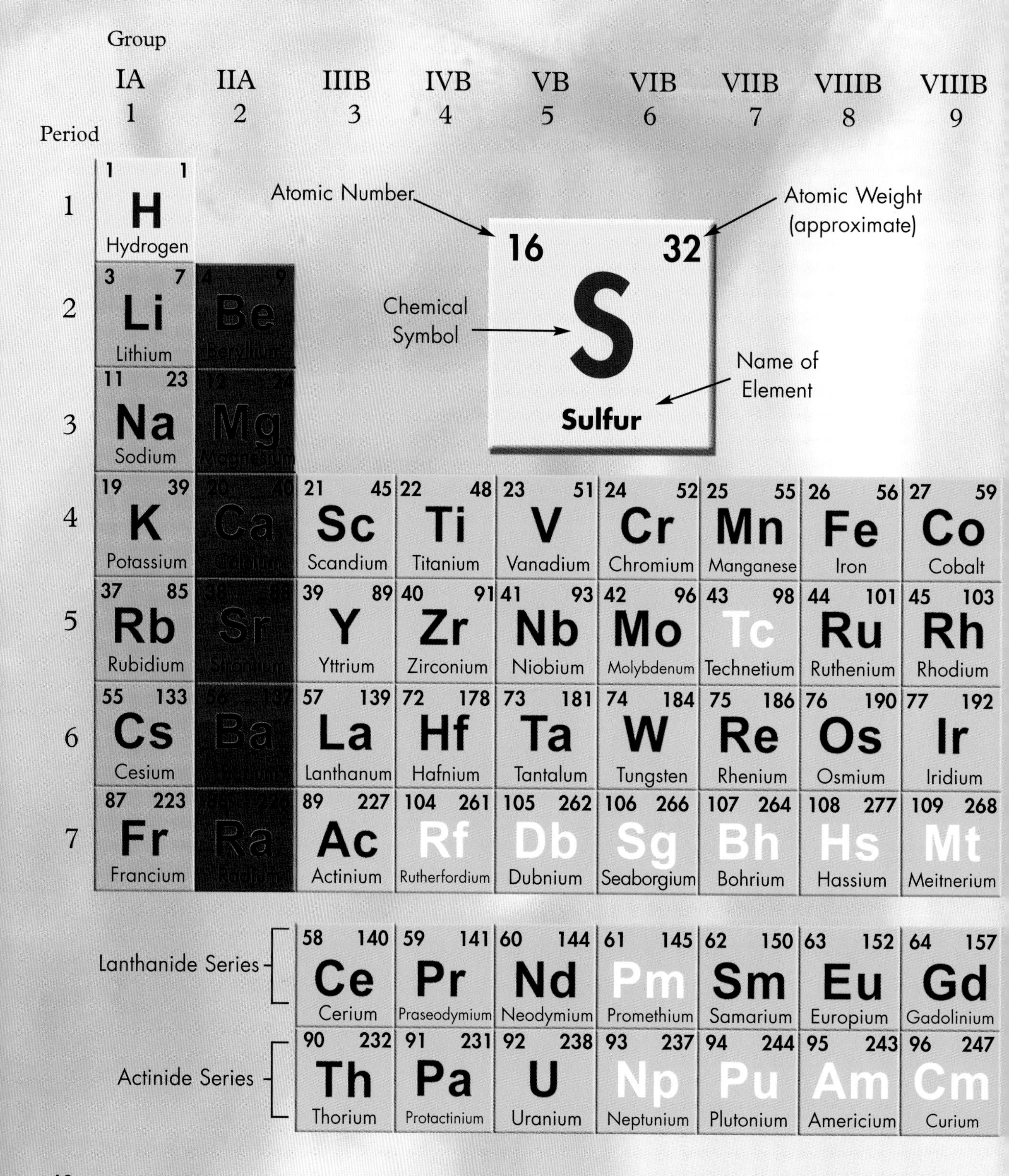

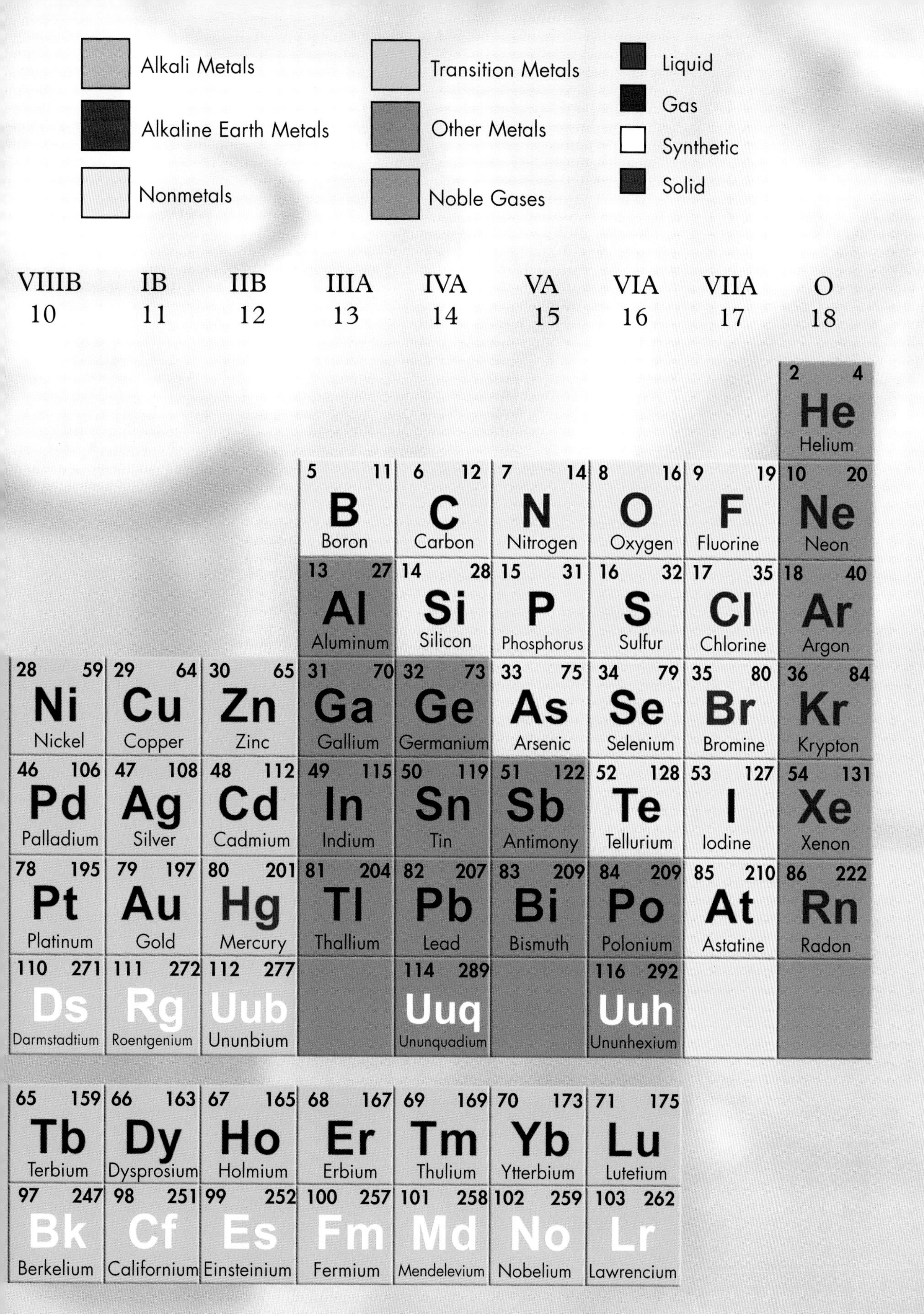
Alkali Metals
Alkaline Earth Metals
Nonmetals
Transition Metals
Other Metals
Noble Gases
Liquid
Gas
Synthetic
Solid
VIIIB 10
IB 11
IIB 12
IIIA 13
IVA 14
VA 15
VIA 16
VIIA 17
O 18
2 4 He Helium
5 11 B Boron
6 12 C Carbon
7 14 N Nitrogen
8 16 O Oxygen
9 19 F Fluorine
10 20 Ne Neon
13 27 Al Aluminum
14 28 Si Silicon
15 31 P Phosphorus
16 32 S Sulfur
17 35 Cl Chlorine
18 40 Ar Argon
28 59 Ni Nickel
29 64 Cu Copper
30 65 Zn Zinc
31 70 Ga Gallium
32 73 Ge Germanium
33 75 As Arsenic
34 79 Se Selenium
35 80 Br Bromine
36 84 Kr Krypton
46 106 Pd Palladium
47 108 Ag Silver
48 112 Cd Cadmium
49 115 In Indium
50 119 Sn Tin
51 122 Sb Antimony
52 128 Te Tellurium
53 127 I Iodine
54 131 Xe Xenon
78 195 Pt Platinum
79 197 Au Gold
80 201 Hg Mercury
81 204 Tl Thallium
82 207 Pb Lead
83 209 Bi Bismuth
84 209 Po Polonium
85 210 At Astatine
86 222 Rn Radon
110 271 Ds Darmstadtium
111 272 Rg Roentgenium
112 277 Uub Ununbium
114 289 Uuq Ununquadium
116 292 Uuh Ununhexium
65 159 Tb Terbium
66 163 Dy Dysprosium
67 165 Ho Holmium
68 167 Er Erbium
69 169 Tm Thulium
70 173 Yb Ytterbium
71 175 Lu Lutetium
97 247 Bk Berkelium
98 251 Cf Californium
99 252 Es Einsteinium
100 257 Fm Fermium
101 258 Md Mendelevium
102 259 No Nobelium
103 262 Lr Lawrencium

Glossary

allotropes Versions of the same element that have different chemical and physical properties.

amorphous Term used to describe substances with a random arrangement of atoms.

atom Tiny particle that makes up all substances. With the exception of hydrogen, which has no neutrons, all stable atoms are made up of protons, neutrons, and electrons.

atomic number Number of protons in the nucleus of an atom.

atomic weight The average mass of an atom of an element.

cellular respiration Conversion of nutrients inside cells into chemical energy that the body can use.

covalent bond Bond between two or more atoms that results from the sharing of electrons.

electromagnetic force Attractive force between objects with opposite charges (e.g., electrons and protons).

electron Negatively charged particle circulating around the nucleus of an atom.

electron shells Spherelike spaces in which electrons circulate around the nucleus of an atom.

ion Atom or group of atoms that have become positively or negatively charged by losing or gaining electrons.

ionic bond Chemical bond in which one atom gives electrons to another, resulting in a positive ion and an negative ion held together by their opposite charges.

monoclinic One of the common crystalline structures into which sulfur atoms naturally arrange themselves.

neutron Subatomic particle with no charge found in the nucleus of an atom.

nucleus Core of an atom, which contains protons and (except in the case of hydrogen) neutrons; the nucleus contains most of the mass of the atom.

orthorhombic Most common crystalline structure into which sulfur atoms naturally arrange themselves.

polymer Large molecule formed by the combination of several smaller, similar molecules, called monomers.

proton Subatomic particle with a positive charge found in the nucleus of an atom.

radioactivity Spontaneous degradation of an element that emits radiation in the form of charged particles in order to reach a stable state.

subatomic particles Particles—including protons, neutrons, and electrons—that comprise atoms.

sulfates Chemical derivatives of sulfuric acid.

valence electrons Electrons traveling in the outermost electron shell of an atom.

vulcanization Process that involves heating rubber and sulfur together in order to harden the rubber.

For More Information

American Chemical Society
1155 16th Street NW
Washington, DC 20036
(800) 227-5558
Web site: http://www.acs.org

International Union of Pure and Applied Chemistry
IUPAC Secretariat
P.O. Box 13757
Research Triangle Park, NC 27709-3757
(919) 485-8700
Web site: http://www.iupac.org

U.S. Environmental Protection Agency (EPA) Headquarters
Ariel Rios Building
1200 Pennsylvania Avenue NW
Washington, DC 20460
(202) 272-0167
Web site: http://www.epa.gov

Web Sites

Due to the changing nature of Internet links, Rosen Publishing has developed an online list of Web sites related to the subject of this book. This site is updated regularly. Please use this link to access the list:

http://www.rosenlinks.com/uept/sulf

For Further Reading

Herr, Norman, and James Cunningham. *Hands-On Chemistry Activities with Real-Life Applications. Easy-to-Use Labs and Demonstrations for Grades 8–12.* New York, NY; Jossey-Bass, 2002.

Keller, Rebecca. *Real Science-4-Kids. Chemistry Level 1.* Albuquerque, NM: Gravitas Publications, Inc. 2005.

Robinson, Tom. *The Everything Kids' Science Experiment Book: Boil Ice, Float Water, Measure Gravity, Challenge the World Around You!* Avon, MA: Adams Media Corporation, 2001.

Strathern, Paul. *Mendelyev's Dream: The Quest for the Elements.* New York, NY: St. Martin's Press, 2001.

Stwertka, Albert. *A Guide to the Elements.* 2nd edition. New York, NY: Oxford University Press, 2002.

Atkins, P. W. *The Periodic Kingdom: A Journey into the Land of the Chemical Elements*. New York, NY: Basic Books, 1997.

Bauman, Robert. *Microbiology*. 2nd ed. San Francisco, CA: Pearson Benjamin Cummings, 2007.

Beatty, Richard. *Sulfur*. New York, NY: Benchmark Books, 2005.

Chang, Raymond. *Chemistry*. 8th ed. New York, NY: McGraw-Hill, 2005.

Emsley, John. *Nature's Building Blocks: An A–Z Guide to the Elements*. New York, NY: Oxford University Press, 2002.

Heiserman, David L. *Exploring Chemical Elements and Their Compounds*. New York, NY: McGraw-Hill, 1991.

Rogers, Kirsteen, et al. *The Usborne Internet-Linked Science Encyclopedia*. London, England: Usborne Publishing Ltd., 2003.

Index

About the Author

Aubrey Stimola has a degree in bioethics from Bard College, where she spent most of her days in the biology and chemistry labs trying to figure out how things work. After two years as the assistant director of public health at a nonprofit organization in Manhattan, she took a research scientist position at the New York State Department of Health. Aubrey is currently a graduate student at Albany Medical College, where she is earning a master's degree in physician assistant sciences. She continues to enjoy writing, which allows her to make the life sciences interesting to young minds.

Photo Credits

Cover, pp. 8, 10, 14, 19, 20, 28, 40–41 illustrations by Tahara Anderson; p. 6 © Richard Treptow/Photo Researchers; p. 7 © Réunion des Musées Nationaux/Art Resource, NY; p. 12 © 2007 periodictable.com; p. 18 © Scala/Art Resource, NY; pp. 23, 25 Shutterstock; Clyde H. Smith/ Peter Arnold Inc.; p. 30 © www.istockphoto.com/Don Wilkie; p. 32 © Prado, Madrid, Spain/The Bridgeman Art Library; p. 35 © Custom Medical Stock Photo; p. 37 © Justin Guariglia/Corbis; p. 39 NASA.

Special thanks to Jenny Ingber, high school chemistry teacher, Region 9 Schools, New York, New York, for her assistance in executing the science experiments illustrated in this book.

Designer: Tahara Anderson; **Editor:** Christopher Roberts
Photo Researcher: Marty Levick